NIGHT SALVAGE

HAUNTING SOCAL'S AUTOMOTIVE GRAVEYARDS

TROY PAIVA

AMERICA
THROUGH TIME®
ADDING COLOR TO AMERICAN HISTORY

America Through Time is an imprint of Fonthill Media LLC
www.through-time.com
office@through-time.com

Published by Arcadia Publishing by arrangement with Fonthill Media LLC
For all general information, please contact Arcadia Publishing:
Telephone: 843-853-2070
Fax: 843-853-0044
E-mail: sales@arcadiapublishing.com
For customer service and orders:
Toll-Free 1-888-313-2665

www.arcadiapublishing.com

First published 2020

ISBN 978-1-63499-244-2

Typeset in Trade Gothic
Printed and bound in England

CONTENTS

ACKNOWLEDGMENTS

This book is dedicated to my Ma: born Stella, then Petea, Lakshmi, and Kali. You had many names, but the singular experience of basking in your bright light made me who I am.

Special thanks to:

David Pearson and the Pearson family.
The memory of Tom Merkel and the Merkel family.
Paul and his amazing collection of junk.

My traveling companions, too numerous to mention. Was that fun or what?

Everyone that took workshops at Pearsonville and Paul's. Your inspiration was my inspiration.

Ryan Baxter and Ron Pinkerton for their inestimable editorial help.

Steve Harper, for inspiring a whole generation of photographers, and my brother Tom, for the influence and encouragement to pursue night photography.

I love Jooolie.

For more fun, visit www.lostamerica.com

INTRODUCTION: THEY ARE US

We are our cars. They define our sense of style, they reflect our station in life. We're born in them, we die in them. *They are us.* Even though they have become interwoven with modern humanity's very DNA, fashions and tastes change quickly as technology accelerates into the future. We may love them, but the fickle reality is the average car only lasts ten years on the road before it ends up in the junkyard. There, it's immediately stripped and crushed, or left to decay on the lot, slowly cannibalized as a parts car.

With the current proliferation of electric vehicles and the soul-crushing prospect of autonomous cars fast approaching, what we require and expect from the automobile continues to change drastically, making vehicles obsolete in only a few years. By comparison, the weirdly sculpted dinosaurs from sixty years ago seem *more* than obsolete—they're like crash-landed spacecraft from another planet.

In the second half of the twentieth century, the American landscape was dotted with junkyards, packed with mid-century iron. It's especially true out west, where the spaces are wide open and land is cheap. Today, these once abundant lots filled with the old, heavy metal are already gone or disappearing fast. Sometimes cities encroach and land values skyrocket, forcing rezoning and clearance. In other cases, obsessive old owners die, and their yard—their life's work—is summarily sold for scrap by their heirs. Every year these "Golden Age of the Automobile" operations become harder to find. They represent a disappearing American legacy.

I have culled the work in this book from three unique and fascinating SoCal collections of junk cars that are going, or already gone. Each reflects something of the personalities and essence of the people who created them. Our first stop on this safari into the surreal is the Pearsonville Junkyard.

EAT

1

PEARSONVILLE

Alongside a bleak and barren stretch of U.S. Route 395 in the California High Desert, Andy and Lucy Pearson founded Pearsonville. It was the dawn of a new decade, the 1960s. The small family had just moved to their recently purchased forty-acre tract in the eastern shadow of the Sierras with plans to start a business servicing motorists. Starting with a drafty two-room shack, they quickly expanded the business into a roadside cafe with an attached apartment.

The run up this stretch of isolated highway was grueling. The relentless grade from the baking SoCal lowlands into the scorching hot upper Mojave took its toll on the cars of the day. With broken-down vehicles being left all over their property, it quickly became clear that Andy and his teenage son, Don, would be spending most of their time towing and repairing the cars of stranded motorists, so they opened a wrecking yard. By the mid-seventies, Pearson's was a thriving operation, quickly becoming one of the largest in the region. What was once nothing but creosote scrub desert fifteen years earlier had become a full-fledged town, proudly featured on roadmaps and regional highway signs.

After two tours of duty in Vietnam as a Navy Seabee, son Don returned to Pearsonville and took on more of the family business. He expanded operations, increasing the size of the towing fleet and opening another yard and recycling facility in Ridgecrest, fifteen miles to the south. Ever the entrepreneur, in 1974 he built a quarter-mile dirt racetrack on the north side of the now sprawling, 4,000-car junkyard. "Pearsonville Raceway" saw wild action on Saturday nights, attracting racers from all over the state.

◀ **EAT Pearsonville**
Pearsonville. 3/17/2008, 11:12 p.m.
Canon 20D, f/5.6, ISO 100. 120 seconds, 4400K WB.
Full moon / white LED flashlight.
Ice clouds burn across the sky, over the derelict cafe on the west side of the highway.

In the eighties, Andy retired and sold out to Don. Splitting his time between the yards in Pearsonville and Ridgecrest, Don's empire flourished and his own son, David, began to participate in the family business. They opened a modern, corporate truck stop and gas station complex on the south edge of Pearsonville in the late nineties, shutting down the original sixties vintage gas station, repair shop, and cafe. By that time, most of the family's wrecking and salvage business had moved to the more centrally located Ridgecrest yard.

When Don suddenly died in 2006, the entire family enterprise fell on David's shoulders. With his plate already full running the Ridgecrest operation, the distant Pearsonville yard became more difficult to justify keeping open. Within a year, thousands of cars were either scrapped or moved to the Ridgecrest yard for storage, while the structures fell into disrepair. Apart from the busy truck stop at the off-ramp, Pearsonville was fast becoming a ghost town.

:::::

This was the state in which I found Pearsonville when I arrived on the scene in March 2008. A fellow night photographer had explored the perimeter of the yard and found a cross country, back way in. He shared his recon with me, adding that it was one of the best junkyards he'd ever seen. On my next trip to the region, I hiked in at midnight, after shooting the nearby ghost restaurant at Dunmovin'. What I found blew my mind. While thousands of cars and trucks had already been removed, there were still approximately 1,000 vehicles inside the fence. The collection that remained was a cross-section of every American marque from the fifties through the seventies, mutely gleaming under the full moon in the chilly spring air. I had stumbled into a mythical elephant's graveyard—forty acres of amazing and rare Detroit iron.

That first night I arrived too late to do much except explore one small section of this enormous site with my jaw hanging open. I cancelled my plans for the rest of that trip. Instead I booked an extra night in my motel in Ridgecrest and hiked into the yard the next evening, shooting and exploring for eight hours. Having only scratched the surface of the site's potential, I spent two of the next three full moon phases driving the 800-mile round trip to the high Mojave, specifically to photograph this amazing collection of vehicles. I slipped in the back way under the cover of darkness a total of six times that summer, a serial trespasser.

:::::

High Sierra Engineering
Pearsonville. 4/24/2010, 11:28 p.m.
Canon 20D, f/5.6, ISO 100. 241 seconds, 5000K WB.
Full moon / white, red and lime-gelled LED flashlight.
Early-fifties Studebaker Champion company car, with a Lark in the background.

After all that sneaking in, I knew I needed to come clean because I wanted to go public with my quickly growing body of Pearsonville work. With a little basic detective work I tracked down the Pearson family. I carefully composed a sincere letter, apologizing for trespassing. I made it clear that I was a harmless, oddball artist, obsessed with shooting and light painting junkyards by the light of the moon, further gushing that this collection was one for the ages.

The third generation owner, David Pearson, ran through the usual range of emotions when he saw the stack of 8 x 10s I sent with that letter: anger, amusement, and then a sense of head-scratching bewilderment at the thought of some nut with a camera roaming the desolate expanses of his family yard by moonlight to make art. I was lucky enough to earn the family's trust and be granted the right to hold night photography workshops on the property.

Teamed with my night shooting partner and photoshop instructor, Joe Reifer, we rented Pearsonville for extended weekends, running workshops a total of seven times between 2008 and 2011. Now we were totally legit. We even had keys to the gate. We'd take full advantage of it too, allowing a dozen photographers to drive into the base camp in the center of the yard, locking them in until 2 or 3 a.m., a night shooting extravaganza.

With afternoon classroom instruction and critique sessions back at the hotel in Ridgecrest, it was a fully immersive experience. While it may sound like the weirdest photography workshop on earth, we were always proud of the fact that half the people that signed up were "repeat offenders," back for more playful camaraderie and endlessly inspiring nights in the most picturesque junkyard in California.

One of the site's best features was the quarter-mile dirt oval racetrack, with its eroding banked turns and rows of splintered bleacher seats. Its last race was held in 2000. With the catchfence poles and lights leaning drunkenly in the subsiding berm and the rotten, half-collapsed press box, the track was a reminder of how quickly the earth reclaims anything man-made when left unattended even for a short time. For people who find beauty and quiet drama in decaying architecture, the raceway is the centerpiece of the Pearsonville experience.

Because the majority of vehicles in the rear section of the yard had been removed, the tightly packed rows of cars you typically find in junkyards were thinned out, leaving a sprinkling of isolated machines with space around them. This gave a unique opportunity to step back and capture individual vehicles as though they were islands in a vast sea of desert, like sand-scoured, sunbaked reefs, teeming with insects and reptilian life. To the west the saw-toothed and treeless southern Sierras tear at the slowly churning UFO-shaped lenticular clouds. To the south, the lights of Ridgecrest twinkle across dry China Lake, fifteen miles away. This backdrop framing the expansive Pearsonville site made it a photographer's dream.

:::::

The Pearsonville you see here no longer exists. In early 2011, hundreds of vehicles were crushed and recycled. The remaining cars were packed tightly into the already crowded northwest corner of the yard and surrounded by a strong, high fence. A forlorn row of trucks and buses continued to sag along the far northern property line, offering a hint of the size of this once vast lot, now reclaimed by coyotes and creosote.

I'm not surprised it happened. Subconsciously, I suppose I was even expecting it. In over thirty years of exploring abandoned places and junkyards, I've seen hundreds of locations disappear. Pearsonville was no different. But the inevitability in knowing that I'd eventually see it go didn't make it any easier.

We stopped doing workshops there. I stayed away for a few years, only to find myself compulsively pulling off the highway and slipping through a crack in the fence in the fall of 2016. While Pearsonville was only a shadow of its former self, it was still like visiting an old friend. I returned a few times in the intervening years to find this windswept place even more ravaged by vandals, even more decrepit and strange. On my last visit, I was stalked half the night by a crazy feral tweaker, banging around in the distance, scrapping parts by moonlight. I kept talking to the silhouette in the darkness in a friendly way, but he never drew closer than about 100 yards. He howled in the distance like a coyote to say good riddance as I finally slipped out the fence in the middle of the night.

With a couple whole cars' worth of parts disappearing to thieves every week, Dave Pearson began the final round of crushing in the summer of 2019. He tells me that by the end of the year, every single car will be gone.

Between 2008 and 2018, I spent thirty-four nights at Pearsonville, always shooting by the light of a full (or nearly full) moon. I've made thousands of hand-lit, minutes-long time exposures, ending up with 576 finished images in my archives. In all those visits, I never ran out of subjects to shoot. I combed every inch of the place and discovered magic everywhere: from its smallest details, to its endless vistas and insanely surreal skies. It will always remain the most fruitful location of my night shooting career.

The Three Eds
Pearsonville. 3/19/2008, 12:53 a.m.
Canon 20D, f/5.6, ISO 100. 241 seconds, 3900K WB.
Full moon, sodium vapor streetlight.

Three 1959 Edsels, lined up as if they were at a Ford dealership in September 1958, snoozing on the far side of the racetrack.

Always Looking North
Pearsonville. 3/18/2011, 12:09 a.m.
Canon 20D, f/5.6, ISO 100. 241 seconds, 5000K WB.
Full moon.
The disintegrating 1922 Seagrave hook-and-ladder firetruck on the racetrack infield, on a stormy winter night.

Dagmars in Heaven ►
Pearsonville. 3/17/2011, 11:15 p.m.
Canon 20D, f/7.1, ISO 100. 181 seconds, 3900K WB.
Full moon / white LED flashlight, purple-gelled strobe.
1956 Cadillac Coupe de Ville, under an encroaching winter storm.

Gnashing Time

Pearsonville. 3/11/2017, 12:55 a.m.
Canon 60D, f/8, ISO 200. 150 seconds, 3800K WB.
Full moon / white, red, and purple Protomachines flashlight.

The Weather Eye ►

Pearsonville. 3/11/2017, 1:11 a.m.
Canon 60D, f/8, ISO 200. 78 seconds, 3800K WB.
Full moon, dark interior / blue, red, and lime Protomachines flashlight.
The peculiar nose and futuristic dash of the pink 1956 Nash Ambassador, settling into the racetrack infield.

The South Forty
Pearsonville. 3/25/2010, 9:29 p.m.
Canon 20D, f/5.6, ISO 100. 241 seconds, 3500K WB.
Full moon.

A partial overview of the back lot at Pearsonville. The tripod was set up at full extension on the back of a flatbed truck.

Haunted Castle
Pearsonville. 3/27/2010, 8:49 p.m.
Canon 20D, f/5.6, ISO 100. 120 seconds, 3500K WB.
Full moon / red, white, and lime-gelled LED flashlight.
Complete 1950 Chrysler Windsor, in the early evening.

◀ **The Weather Channel**
Pearsonville. 4/25/2010, 11:36 p.m.
Canon 20D, f/5.6, ISO 100. 120 seconds, 3900K WB.
Full moon / red, white, and lime-gelled LED flashlight.
TV and tumbleweeds in the trunk of a 1955 Chevy, lurking in the yard's back lot.

Long in the Roof
Pearsonville. 7/16/2008, 12:21 a.m.
Canon 20D, f/5.6, ISO 100. 120 seconds, 3850K WB.
Full moon / red and blue-gelled LED flashlight.

An extremely rare Checker Marathon Aerobus limousine. They had two of them.

The Bucket Express

Pearsonville. 3/18/2008, 12:36 a.m.

Canon 20D, f/5.6, ISO 100. 120 seconds, 3600K WB.

Full moon / red-gelled strobe flash.

Fifties vintage GM "Old Look" transit bus. Shot outside the fence on my first visit. It was moved inside the perimeter by summer.

GP '73 ►

Pearsonville. 7/14/2008, 10:18 p.m.

Canon 20D, f/5.6, ISO 100. 181 seconds, 3850K WB.

Full moon / white and lime-gelled LED Flashlight.

My family bought a 1973 Grand Prix when I was thirteen. One of my favorite GP adventures was when my mom let me drive down 395, all night, on my learner's permit while she slept. Pearsonville was closed for gas and we rolled into Ridgecrest on fumes at 4 a.m.

Race Control
Pearsonville. 7/15/2008, 11:38 p.m.
Canon 20D, f/5.6, ISO 100. 120 seconds, 3950K WB.
Full moon, dark interior / red and lime-gelled LED flashlight.

Overlooking turn one, from the rickety, guano-covered press box of "Pearsonville Raceway", the quarter-mile dirt track erodes in the middle of the yard.

◄ **Crumbling Bakelite**
Pearsonville. 3/06/2009, 9:33 p.m.
Canon 20D, f/5.6, ISO 100. 199 seconds, 3500K WB.
Full moon, dark interior / white, red, and lime-gelled LED flashlight.
A 1941 Packard Clipper near the back fence. This car disappeared in the 2011 crush. I doubt it was saved.

Waiting for Westergard
Pearsonville. 3/06/2009, 9:54 p.m.
Canon 20D, f/5.6, ISO 100. 210 seconds, 3100K WB.
Full moon / white, red, and lime-gelled LED flashlight.

Mollusk Abuse
Pearsonville. 5/27/2018, 1:46 a.m.
Canon 60D, f/8, ISO 200. 180 seconds, 4500K WB.
Full moon / blue, red, and purple Protomachines flashlight.
1956 Cadillac RV conversion affectionately known as "The Mollusk" or "Caddyshack." It's been dragged around the lot over the years, and worse for wear after every move. Of all the vehicles at Pearsonville, it's the one I've photographed the most.

◀ **Series 62**
Pearsonville. 3/18/2008, 12:51 a.m.
Canon 20D, f/5.6, ISO 100. 120 seconds, 5000K WB.
Full moon / red-gelled strobe flash, white, red, and blue-gelled LED flashlight.

Ghost Coach

Pearsonville. 10/14/2016, 9:26 p.m.
Canon 60D, f/8, ISO 200. 98 seconds, 3800K WB.
Full moon / white and lime Protomachines flashlight.

Red Velvet in a Purple Velvet World
Pearsonville. 10/14/2016, 9:43 p.m.
Canon 60D, f/8, ISO 200. 180 seconds, 3900K WB.
Full moon, dark interior / purple and lime Protomachines flashlight.
Inside and out of a fifties GM city bus, converted to RV duty in the sixties, parked ever since.

Constellation Montclair
Pearsonville. 4/23/2010, 10:01 p.m.
Canon 20D, f/5.6, ISO 100. 241 seconds, 3500K WB.
Full moon and sodium vapor streetlight / red LED flashlight, laser pointer.

One of my workshop students brought a laser pointer with a lens that scattered the beam into hundreds of points of light. Experimental amusement ensued, all over this 1965 Mercury Montclair.

Pearson de Ville
Pearsonville. 5/27/2018, 11:24 p.m.
Canon 60D, f/7.1, ISO 200. 180 seconds, 4650K WB.
Full moon / white and red Protomachines flashlight.
1961 Cadillac Sedan de Ville with its unusual six pack of lights.

Patrician Cathedral

Pearsonville. 3/11/2017, 11:28 p.m.

Canon 60D, f/8, ISO 200. 64 seconds, 5000K WB.

Full moon / white, red, and purple Protomachines flashlight.

The streamlined apse of a 1956 Packard Patrician.

Joker Taffy
Pearsonville. 10/15/2016, 12:48 a.m.
Canon 60D, f/8, ISO 200. 180 seconds, 3800K WB.
Full moon, partial cloud cover / white, green, and red Protomachines flashlight.
Scowling fifty-nine Buick, with super stretchy, long exposure cloud porn.

The Periwinkle Spaceman
Pearsonville. 9/24/2010, 1:04 a.m.
Canon 20D, f/8, ISO 200. 241 seconds, 3900K WB.
Full moon / red and purple-gelled LED flashlight.
The strangely organic 1960 Dodge Dart Seneca.

Vista Malaise

Pearsonville. 6/24/2018, 11:22 p.m.

Canon 6D Mk II, f/8, ISO 200. 242 seconds, 4200K WB.

Full moon / white, purple, red, and blue Protomachines flashlight.

Crappy mid-seventies Oldsmobile VistaCruiser wagons (crappy if for no other reason than there's just a small sunroof instead of the amazing panoramic roof glass on the previous models), its shelf-paper wood paneling bleached and cracked by the desert sun.

The Black Hole

Pearsonville. 7/15/2008, 12:09 a.m.

Canon 20D, f/5.6, ISO 100. 120 seconds, 3850K WB.

Full moon / white LED flashlight.

Looking up, into a 1966 Ford, topping a pile of cars, on a brutally hot and muggy night.

The God of Light
Pearsonville. 5/27/2018, 9:19 p.m.
Canon 60D, f/8, ISO 200. 78 seconds, 5550K WB.
Full moon / white, red, and purple Protomachines flashlight.
Mid-seventies Buick Apollo. Super rare ugly duckling of a car, almost entirely forgotten by 2020. There were two at Pearson's.

Fiat Amputata ▶
Pearsonville. 10/04/2009, 9:40 p.m.
Canon 20D, f/5.6, ISO 100. 241 seconds, 3900K WB.
Full moon / white and purple-gelled LED flashlight.
Fiat 128 front clip, hidden in the northeast corner of the yard. It was the only Italian car left in Pearsonville.

Heavy Seas
Pearsonville. 3/08/2009, 12:31 a.m.
Canon 20D, f/5.6, ISO 100. 120 seconds, 3900K WB.
Full moon / white and lime-gelled LED flashlight.

Grandpa's early-sixties Ford pickup, protected by Smith and Wesson.

XL
Pearsonville. 10/14/2016, 11:47 p.m.
Canon 60D, f/8, ISO 200. 75 seconds, 3800K WB.
Full moon, partial cloud cover / white Protomachines flashlight.
A 1968 Ford growls, fisheyed along the front wall.

Monte Carlo Moonrise ►
Pearsonville. 10/04/2009, 7:42 p.m.
Canon 20D, f/5.6, ISO 100. 120 seconds, 5000K WB.
Full moon / red, green, and lime-gelled LED flashlight.
First generation Chevy Monte Carlo. The exposure was started as the full moon began to rise and it literally rose into the shot as I lit the scene.

Thunderbirds Are Go!

Pearsonville. 4/23/2010, 11:23 p.m.

Canon 20D, f/5.6, ISO 100. 241 seconds, 5000K WB.

Full moon / red, white, and purple-gelled LED flashlight.

Thunderbird Row, a flock of over a dozen T-birds from five generations in the front lot.

The Skinny on the Mini

Pearsonville. 10/14/2016, 11:03 p.m.

Canon 60D, f/8, ISO 200. 116 seconds, 3800K WB.

Full moon, partial cloud cover / red, green, and purple Protomachines flashlight.

The remains of the long-closed Mini Mart and pump islands.

Pterodactyl Saddle
Pearsonville. 5/27/2018, 10:04 p.m.
Canon 60D, f/8, ISO 200. 180 seconds, 4400K WB.
Full moon / red and purple Protomachines flashlight.
The biggest, fattest, mid-seventies T-bird ever, slowly decaying on Thunderbird Row.

Barracuda Sarcophagus ▶
Pearsonville. 5/27/2018, 10:57 p.m.
Canon 60D, f/8, ISO 200. 134 seconds, 3600K WB.
Full moon / lime, red, and blue Protomachines flashlight.
Basket case mid-sixties fastback Plymouth, with rotten and dusty custom upholstery, haunting the front lot.

Mandibles
Pearsonville. 3/12/2017, 1:20 a.m.
Canon 60D, f/8, ISO 200. 180 seconds, 5500K WB.
Full moon / white, lime, blue, and purple Protomachines flashlight.

1961 Plymouth Belvedere, generally considered one of the ugliest cars of all time. I love its bug-face through the fisheye and that quirky two-degree cant to the headlights. It's so subtle that when I pointed it out to a guy who owned one, he'd never noticed.

El Matador Thinks ►
Pearsonville. 3/28/2010, 9:31 p.m.
Canon 20D, f/5.6, ISO 100. 241 seconds, 3400K WB.
Full moon / red LED, white xenon flashlights.
A contemplative 1974 AMC Matador, under high Sierra lenticular clouds.

The Talking Tree
Pearsonville. 10/01/2009, 10:00 p.m.
Canon 20D, f/5.6, ISO 100. 241 seconds, 3900K WB.
Full moon / red LED, white xenon flashlights.
The one of a kind DIA•LOG truck, a postwar International ambulance, converted for wire and cable laying duty.

The Night Rider
Pearsonville. 10/14/2016, 10:39 p.m.
Canon 60D, f/8, ISO 200. 146 seconds, 5000K WB.
Full moon, partial cloud cover / white, green, and purple Protomachines flashlight.
Knight Rider 2010 TV-movie prop car *Hannah*, decaying along the front fence. Postapocalyptic sci-fi wreckage at the Trading Post.

Magnum Lenticularis
Pearsonville. 10/14/2016, 8:13 p.m.
Canon 60D, f/8, ISO 200. 180 seconds, 3900K WB.
Full moon, partial cloud cover / red and lime Protomachines flashlight.
Third-generation Dodge Charger scraps, fisheyed, under liquid skies.

Ville Phantome

Pearsonville. 9/25/2010, 1:24 a.m.

Canon 20D, f/8, ISO 200. 241 seconds, 3900K WB.

Full moon / red and purple-gelled strobe.

Pretty sure it was while shooting this image when I lost my strobe flash in this walled section of the yard. Before I had a chance to return to look for it, the yard was cleaned out and every car you see in this image disappeared, along with my Canon 430EX.

Hang Five ►
Pearsonville. 9/21/2010, 11:45 p.m.
Canon 20D, f/8, ISO 200. 241 seconds, 3500K WB.
Full moon / white and red LED flashlight.
Shooting this 1962 Pontiac Catalina wreck was all about the gas pedal.

Floating Headlights
Pearsonville. 3/18/2008, 1:14 a.m.
Canon 20D, f/5.6, ISO 100. 120 seconds, 3750K WB.
Full moon, sodium vapor streetlights / red-gelled strobe flash, white, and lime-gelled LED flashlight.
1963 Chrysler Imperial, chin first, hunkered under the creosote and racing clouds.

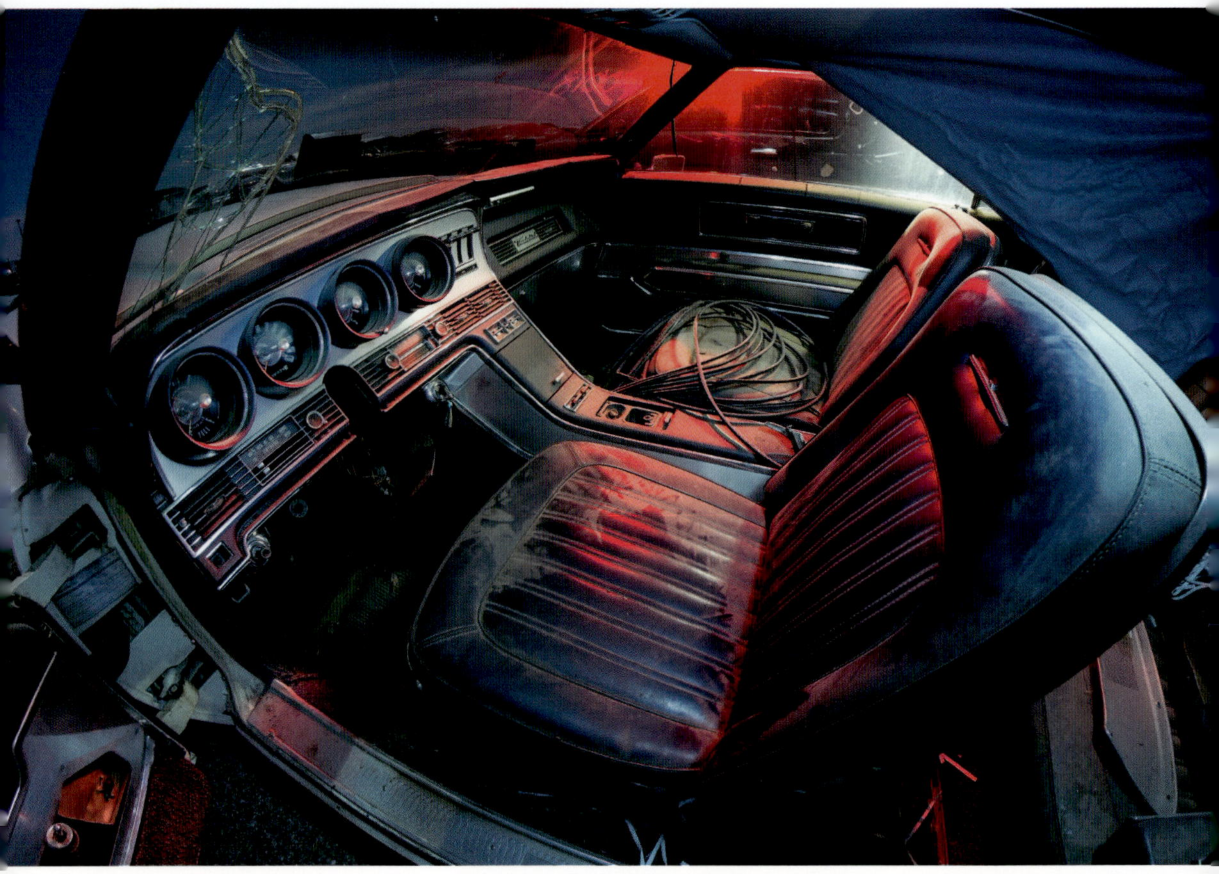

Glamorbird Cockpit
Pearsonville. 3/12/2017, 12:26 a.m.
Canon 60D, f/8, ISO 200. 89 seconds, 3800K WB.
Full moon, dark interior / white and red Protomachines flashlight.
Fisheyed in the cockpit of a 1969 Thunderbird. Even more sci-fi without the steering wheel, it's as though you steer by thought.

ACME Strobe Stripe ►
Pearsonville. 5/26/2018, 11:54 p.m.
Canon 60D, f/8, ISO 200. 100 seconds, 4300K WB.
Full moon / blue and yellow Protomachines flashlight.
Fake Road Runner stripe on a crashed seventy-two Plymouth Satellite.

2

THE CUYAMA CAR GARDEN

I'd wanted to visit Tom Merkel's Car Garden for years without even knowing its name or if it really existed. It's one of those legendary places people whisper about, but no one knows (or says) where it is. I'd heard the tales more than once in bull sessions with junkyard guys: "There's a motherlode of cars, lost up some remote SoCal canyon." A 2005 article in *Car and Driver* magazine called "Midnight in the Garden of Eldorados and E-Types" was confirmation in the mainstream media, proving the rumors true. The magazine's editors were allowed to run the article only if they didn't disclose the location. Fascinated, I filed it away as one more "wish list" location and promptly forgot about it for ten years.

Shortly after Merkel died in 2016, stories and videos about the Garden began to surface on the web. The location had leaked and was now even marked "Tom Merkel's Cuyama Car Garden" on Google Maps. The swollen river of cars snaking up the canyon was impressive in the satellite view. I immediately dropped all my other night shooting plans and made the five-hour drive to one of the most remote corners of SoCal for the September 2017 full moon, with nothing but a pin drop on a map and no idea what would happen.

At the turn off from the highway, the dirt road was gated, marked only by a few more "No Trespassing" signs than usual. I parked by a closed roadhouse at the junction, hopped the gate and hiked in the last mile, past about forty more signs and over a couple more gates. I never saw a soul. Once I arrived onsite, I spent a few hours staggering from one car-jammed clearing to another in the hot sun. The slow reveal of treasure after treasure was overwhelming. As the trail twisted ever deeper into the canyon, it dawned on me that this is what it must feel like to find the Holy Grail … of junk car collections.

:::::

Tom Merkel was a man ahead of his time, a visionary artist. He grew up in a house surrounded by art. His mother's pioneering collection of African Art—the Katherine White Collection—is considered one of the best in the world, and is now the pride of the Seattle Art Museum. Clearly, this early exposure to ancient primitive art and anthropology stuck with him. Combined with his earthy hippy nature, it makes sense he'd live in a remote cabin in the woods surrounded by talismanic piles of rusty twentieth-century artifacts.

Merkel was also an heir to Cleveland's White Motor Company, famous for its early steam cars, Depression-era tour buses, and postwar semi-trucks. His grandfather, Walter White, even raced in the 1905 Vanderbilt Cup race in a White Steamer. With the White family oil running through his veins, it's no surprise that Tom started collecting cars as a teenager. But his passion was old junkers—the quirkier, the better.

Stalking salvage yards surrounding his Santa Barbara home in the early 1980s, he found $200 could rescue drivable, but tired old Packards, Lincolns, and Cadillacs, just moments from the crusher. He obsessively bought whole fleets of worn-out trucks and company cars. He cleaned out police impound lots and combed local Penny Saver ads for $50 jalopies. Every vehicle was unique, still loaded with the cultural and anthropological artifacts of their previous owners. He understood this hidden significance in his exploding collection, evidenced by his handwritten manifesto sent to *Car and Driver* magazine: "[The Car Garden is] essentially a gallery of automotive self-portraits, where a single copy … becomes as easily recognizable and unique in personality as the owners were themselves. The cars are as if they were pages from hundreds of family albums."

He had vehicles stashed in lots and garages all over town, but it quickly got out of hand. In 1984, he purchased eighty acres in a remote canyon south of Cuyama and began to shuttle cars to their new home. Given the winding, mountainous three-hour drive from Santa Barbara, the process was slow. If the vehicle could make the trip, it was driven one last time, to its final resting place at the Garden. If not, he brought them a couple at a time on the back of, or towed behind, his tired 1962 Ford flatbed truck. It took more than ten years to drag everything over the mountain and up the canyon to create his masterpiece.

:::::

Nothing could have prepared me for the scale of the Cuyama Car Garden. I barely scratched the surface in my two afternoons and nights of wandering around—gobsmacked by the more than 1,100 trucks, buses, cars and other petro-related objects, scattered over this wilderness landscape, miles from nowhere. While it may have appeared random to some, there was obvious and playful organization going on.

Your Instructions

Cuyama Car Garden. 9/4/2017, 9:32 p.m.

Canon 60D, f/8, ISO 200. 177 seconds, 3900 WB.

Full moon / white Protomachines flashlight.

The sign at the entrance reads: "The Cuyama Historical Car Garden: a creative preservation project based on the oriental rock garden concept of balance between form and nature. A memorial to the men of machines, and the machines of men, a sanctuary of form, time and witness. Please be respectful. By invitation only - no parts sold. Estimated completion - 1995."

In one clearing, a simulated *Blues Brothers* chase, with a fleet of authentic blue and whites, was lined up behind the Bluesmobile. Hidden down one of the side trails were the sixties micro-Fiats of "Little Italy" and the "Mercury Offramp to Oblivion," an amazing pack of rare late-fifties Big M Mercs. They were all straight and complete, with butts still in the ashtrays and registration, maps, and oil change receipts still in the glove compartments. The centerpiece tying all these tableaus together was "The Lost Highway," a four-car wide, simulated traffic jam, that wound its way up the canyon, bumper-to-bumper, for a mile. Merkel called it "a deliberate ruin, a cultural time capsule of sorts, a celebration of Southern California auto culture intended for the study and enjoyment of generations to come."

Always secretive about the location, Merkel's intention may have been to let the whole thing vanish into the mountainside to be found by archaeologists from the distant future. He's also quoted as saying, "The work in progress is basically the twentieth-century equivalent of the Chinese clay soldiers"—referring to the 2,200-year-old Terracotta Army unearthed in China in 1974.

Now, twenty-five years on, many of the cars were disappearing into the dense banks of sage and mesquite, while others were partially consumed in mud flows and washes, as if they were encased in sediment at the bottom of the sea. Whatever Merkel's intentions, with cars already well on their way to completely disappearing into the terrain, I found it all achingly, tragically poetic and beautiful.

The kaleidoscopic spectrum of brands and models that ran from the mundane to one-of-a-kind rarities was unparalleled. Whatever your personal obsession, the Garden had it covered: Postwar European micro-cars? City buses? Gas pumps? Sports cars? Muscle cars? Luxury marques? Garbage trucks? They were all there. Me? Hearses. I couldn't believe all the cool hearses! Did I see *Harold and Maude* too many times when I was a kid? Maybe, but for whatever reason, I ended up making an inordinate number of hearse pictures. The centerpiece for me was the 1959 Cadillac on page 77 that once belonged to star-crossed actor Brandon Lee.

There were so many things I missed shooting: a ghostly fleet of white 1958 Chevy sedans and wagons, loaner cars from a friendlier bygone era, the fading name of a local airport, stenciled on the doors. At least ten of them, all in a cluster. I'm still kicking myself for skipping the BMW Isetta or Messerschmitt, or even just seeing the customized short wheelbase 1950 Cadillac hearse with my own eyeballs, instead of in another visitor's snapshots months later.

:::::

I explored all afternoon with only my phone camera. Hiking back out, wondering if sneaking in tonight would get me shot, I couldn't believe my luck when I happened to meet the current caretaker at the gate. Using the old heart-on-my-sleeve, passionate artist technique, I sold him on allowing me to shoot for two nights. In our conversation, I was shocked to hear that he was tasked with the liquidation of the entire collection. It was already underway and most of the cars had been accounted for. This sad news immediately redoubled my determination for making the most of this trip. I might never get another chance.

With no time to get back to a town for a meal, or book a motel, I returned an hour later at dusk and hiked in again with all my gear. Shooting nonstop until after midnight, when all my batteries were flat and I could barely stand up anymore, I hiked back to the car and drove to a turnout at the top of a nearby peak. There I curled into a ball in the back seat, finally lulling myself to sleep for a couple hours by muttering the mantra, "I'm getting too old for this shit," to myself repeatedly. By 10 a.m., I'd found a cheap motel in Maricopa where I flopped for the day, recharging.

When I showed up a couple hours before dark on the second day, the gate was locked. Undeterred and reenergized after snoozing in the motel all afternoon, I hiked in one more time. I made my way into the deepest reaches of the Garden. More dense and eclectic than the front sections, the truly remarkable cars were tucked way back in there.

Around 10 p.m. on the second night, a figure approached, shining a flashlight. I called out using the caretaker's name: "Hey, is that you?" The silhouette said, "No. I live here. You're trespassing." Angry. Drunk. Looking for trouble. I confidently replied, "I have permission. I'm just taking pictures." To which he sarcastically slurred, "Yeah sure. At night? Bullshit. Let me see the inside of your pack." I did. No parts or tools in there. "Look, I'm even packing out my empty drink bottles."

No matter what I said, he was having none of it. I cursed myself for not carrying one of my earlier books in the pack. I showed him thumbnails on the back of my camera, hoping to make him understand. He responded by calling me a "stupid tweaker." He was convinced it was all just a ruse to perform a little midnight auto supply.

I think when I said, "The actual caretaker gave me permission to be here, I don't even know who you are to say I can't be here" was when it started to go off the rails. When I said to him, "Oh hey, please don't step there, you're making footprints," he proceeded to throw me out on the spot. "I don't like your attitude, I'm gonna check your bag again by the gate on your way out. Finish this stupid picture and be at the gate in ten minutes!" He then staggered off, into the darkness.

Dumbfounded, I stood there frozen in my tracks. Now what? Naturally, I waited a few minutes and then, with a shrug, wandered deeper into the canyon, shooting until three, when I ghosted my way down the trail and out without a sound. I heard some yelling in the distance about an hour after I'd ditched him, so who knows what transpired between the caretaker and this other guy. It was the sketchiest confrontation I've had in years.

With the hope of a return visit, I hammered the caretaker with many texts and photos, but he never replied. His voicemail box was always full. It didn't take long to come to the realization that the Car Garden, once one of the best kept secrets of the car world, had now become one of the worst. Collectors and brokers from all over the country were bidding on cars and hauling them out by the truckload through the fall and winter. My friend John at Big M Auto Salvage dug out a couple of late-fifties MOPAR cars in January 2018 and said there were only "a couple hundred" cars left. Exploring it with Google Maps today reveals less than 100 cars, still scattered up and down the now mostly bulldozed canyon.

It's already gone. Merkel's massive art experiment, meant for some future generation to discover, barely outlived him a year. It's every artist's worst nightmare: all that work, all that effort, and almost no one ever saw it or knew it even existed. I wish I could have met him in the nineties and had been already night shooting there for years. I like to think that we were kindred spirits and he would have enjoyed my efforts to capture and preserve his vision.

While I could have spent fifty nights there, I only had two. The shots here depict individual cars which cannot begin to capture the breadth and scale of the entire project, now the stuff of legends. Looking back, the whole weekend feels like a dream.

Stepside Moonrise
Cuyama Car Garden. 9/5/2017, 8:00 p.m.
Canon 60D, f/8, ISO 200. 42 seconds, 3900 WB.
Full moon / white Protomachines flashlight.
1950s Studebaker pickup, burned in a brushfire and encased in a mudflow.

Chin in the Dirt ►
Cuyama Car Garden. 9/5/2017, 7:51 p.m.
Canon 60D, f/8, ISO 200. 5 seconds, 3800 WB.
Deep shade / white Protomachines flashlight.
1950 Buick hiding in the sage.

When the Waters Recede
Cuyama Car Garden. 9/5/2017, 9:20 p.m.
Canon 60D, f/8, ISO 200. 170 seconds, 3800 WB.
Full moon / white and blue Protomachines flashlight.
Mudflow trapped 1960 Pontiac Safari and 1959 Cadillac Coupe de Ville, forgotten up the canyon in the name of art.

◄ **The Discount Hearse**
Cuyama Car Garden. 9/6/2017, 2:12 a.m.
Canon 60D, f/8, ISO 200. 161 seconds, 3950 WB.
Full moon / white, red, and green Protomachines flashlight.
1963 Buick Flxible hearse, halfway up the south side of the canyon.

ROVER
BS 4262

8101
Cuyama Car Garden. 9/6/2017, 1:28 a.m.
Canon 60D, f/8, ISO 200. 241 seconds, 3800 WB.
Full moon / white Protomachines flashlight.
Shining a flashlight at a firetruck in the middle of the night, trapped in the mudflows.

◀ **The Viking's Head**
Cuyama Car Garden. 9/6/2017, 2:26 a.m.
Canon 60D, f/8, ISO 200. 16 seconds, 3900 WB.
Deep shade / white Protomachines flashlight.
Rover P4 in a dark spot under the oak trees lining the canyon.

MOTOR
HARLEY-DAVIDSON
CYCLES
2EXK948

Crawling from the Tarpit
Cuyama Car Garden. 9/5/2017, 10:19 p.m.
Canon 60D, f/8, ISO 200. 241 seconds, 3800 WB.
Full moon / blue and purple Protomachines flashlight.

◀ **Harold Puh-leez!**
Cuyama Car Garden. 9/5/2017, 10:39 p.m.
Canon 60D, f/8, ISO 200. 64 seconds, 5000 WB.
Full moon / white, red, and purple Protomachines flashlight.
The 1959 Miller Meteor Panoramic Cadillac hearse, pitched and frozen in a mudflow to its frame. The backstory on this car is that it belonged to actor Brandon Lee, who sold it to Merkel when he became tired of being stopped by the cops all the time.

Wicker and Weeds
Cuyama Car Garden. 9/5/2017, 9:08 p.m.
Canon 60D, f/8, ISO 200. 77 seconds, 3800 WB.
Full moon / white, red, and green Protomachines flashlight.
1950 Cadillac hearse, with most of its thick black lacquer scorched off by the sun. The ancient wicker wheelchair rotting inside might be its best feature.

◀ **The Dismal Trade**
Cuyama Car Garden. 9/5/2017, 8:47 p.m.
Canon 60D, f/8, ISO 200. 241 seconds, 3800 WB.
Full moon / white and green Protomachines flashlight.

The Continental Shelf
Cuyama Car Garden. 9/5/2017, 11:40 p.m.
Canon 60D, f/8, ISO 200. 241 seconds, 3800 WB.
Full moon / white Protomachines flashlight.
Suicide door Connie trapped in a mudflow.

Bonsai de Ville ►
Cuyama Car Garden. 9/5/2017, 11:55 p.m.
Canon 60D, f/8, ISO 200. 151 seconds, 5000 WB.
Full moon / white, purple, and green Protomachines flashlight.
A dwarf tree grows close, blocking the trunk of a 1953 Cadillac.

Save the Seals
Cuyama Car Garden. 9/4/2017, 11:36 p.m.
Canon 60D, f/8, ISO 200. 22 seconds, 3400 WB.
Deep shade / white and red Protomachines flashlight.
A "Save the Seals" bumper sticker on a 1974 Mazda RX-4, with an early Wankel engine, notorious for blowing its engine seals.

◀ **Six Feet Under the Thistle Patch**
Cuyama Car Garden. 9/5/2017, 8:33 p.m.
Canon 60D, f/8, ISO 200. 171 seconds, 3500 WB.
Full moon / white, red, and blue Protomachines flashlight.
Rustbucket 1971 Miller Meteor conversion, sinking into a deep thorny patch.

The Pink Saloon
Cuyama Car Garden. 9/5/2017, 11:25 p.m.
Canon 60D, f/8, ISO 200. 195 seconds, 5000 WB.
Full moon / white, red, and purple Protomachines flashlight.
1950s Austin A30, a funky, little upright car.

◄ **Grandfather Clock**
Cuyama Car Garden. 9/6/2017, 1:06 a.m.
Canon 60D, f/8, ISO 200. 241 seconds, 3800 WB.
Full moon / white and purple Protomachines flashlight.
1954 Ford and a Neptune 855 pump with a Seaside Silver Gull graphic, deep in a thicket. Dozens of old pumps lined the Garden's far edges.

Triumph of the Chaparral
Cuyama Car Garden. 9/5/2017, 10:49 p.m.
Canon 60D, f/8, ISO 200. 241 seconds, 3800 WB.
Full moon / white Protomachines flashlight.
Burned and half-buried, the Car Garden was hard on this Triumph TR3.

◀ **Riding the Sage Surf**
Cuyama Car Garden. 9/4/2017, 10:51 p.m.
Canon 60D, f/8, ISO 200. 37 seconds, 3800 WB.
Deep shade / white, red, and lime Protomachines flashlight.
1961 Cadillac, settling on a spur trail.

Deadfall de Ville
Cuyama Car Garden. 9/5/2017, 9:54 p.m.
Canon 60D, f/8, ISO 200. 241 seconds, 3800 WB.
Full moon / purple and red Protomachines flashlight.
1959 Cadillac Sedan de Ville, disappearing into the countryside.

Aronde Bend ►
Cuyama Car Garden. 9/5/2017, 10:58 p.m.
Canon 60D, f/8, ISO 200. 42 seconds, 3800 WB.
Deep shade / white and blue Protomachines flashlight.
1950s Simca Aronde, far from home.

ARONDE
CALIFORNIA
IJH 935

Bureau of Prisons
Cuyama Car Garden. 9/5/2017, 11:48 p.m.
Canon 60D, f/8, ISO 200. 241 seconds, 3800 WB.
Full moon / white Protomachines flashlight.
Dappled moonlight on an early-sixties crew cab Power Wagon, a former prison truck.

◄ **Hippy Heaven**
Cuyama Car Garden. 9/6/2017, 2:47 a.m.
Canon 60D, f/8, ISO 200. 176 seconds, 5000 WB.
Full moon / white and purple Protomachines flashlight.
I hiked back and forth past this homemade camper on a VW Kombi, in a clearing mostly by itself, several times in my two nights, saving it for the last shot of the weekend.

3

PAUL'S JUNKYARD

It's the July 2019 full moon and I'm at my old pal Paul's High Desert junkyard. I've been night shooting here almost forty times under every kind of conditions imaginable: from blast furnace heat, to that howling frozen wind that only happens in the Mojave. I've been there with big groups of students—herding cats on workshop nights—but I usually come alone or with one other photographer.

Perhaps I was being cavalier or complacent on this beautiful summer night, but as I finished shooting the *Cobb* movie prop Imperial, my feet got tangled, throwing me onto those rusty girders visible in the image. As I fell, I rotated my body to protect my expensive flashlight, landing squarely on my right shoulder. Excruciatingly, I felt the muscle pop right off the bone. I rushed home to the doctor the next day, the MRI revealing a 100% tear. In early August I had rotator cuff surgery. It's now five maddening weeks into a six-month recovery. I still can't really move it, but at least I can type without pain now.

My introduction to Paul's High Desert salvage operation took place in March 2002 when I stumbled on it while trolling around on some old backroads in the Mojave. While the city buses, searchlights, and other curious mechanical artifacts stashed along that big steel wall slowed me down, the partially open rolling gate invited me in. Paul, a burly cinderblock of a man with a perpetual grin under his thick and sandy trucker mustache, went agog when I showed him my sample 35mm slides of my earlier night photography (yes, that's what I carried in that seemingly ancient, pre-digital era). I watched him puff with pride at the idea that someone wanted to turn *his* joint into an art project.

And what a joint it is! Paul's father started a modest salvage operation there in 1969. He started working for his old man as a teenaged yard hand, recycling Mojave ranch scrap in the seventies. By the early nineties, he'd bought his dad out and ramped up operations.

One of Paul's poker buddies had a business building and leasing vehicular props for Hollywood. Paul's yard became an overflow spot to stash all kinds of objects once

Shoved by Ty Cobb's Ghost
Paul's Junkyard. 7/13/2019, 10:54 p.m.
Canon 6D Mk II, f/8, ISO 200. 130 seconds, 3800 WB.
Full moon / lime and purple Protomachines flashlight.
One of two prop Imperials for the 1994 movie *Cobb*, in that hard-to-get-to corner of Paul's. This was the last shot before I fell, tearing my rotator cuff. I landed right on those girders.

filming ended. Vintage cars, trucks and buses—as well as NYC subway cars, jet aircraft and helicopters—have all been stored inside the fence at one time or another. Some vehicles were trucked back and forth to Los Angeles, to be repainted and reused in different productions. Many were multiple copies of the same vehicle, with various levels of crash damage from chases and stunts. Some were total wrecks—mangled, burned, or pocked with faked bullet holes—the stunt-driver safety-cell still intact inside.

They could sit rotting for years before Paul would buy them for scrap. If he liked them, he'd keep them around for a while, sometimes indefinitely, but most of the props were destined to meet their end in The Pit, in the center of the yard. There, he'd use massive crushers and shredders to reduce the piles of vehicles and other machines into recyclable pieces no bigger than your fist. From the air it looks like a circling black whirlpool, slowly absorbing everything in its orbit.

Paul also combed police and military vehicle auctions for the unusual and strange. Mix that with his thing for giant stationary engines, rusty farm equipment, and fire trucks, and you'll understand what it's like to wander around at Paul's.

After the loss of Pearsonville as our workshop site in 2011, my first instinct was to contact Paul. Joe and I pulled into the yard that crisp January afternoon, just as Paul was closing a deal, buying or selling some weird treasure. Even though it had been nearly ten years, Paul remembered me and grinned when I gave him a copy of my first book, *Lost America,* featuring several images shot there in 2002.

He said "Yes!" to our workshop idea instantly, and we ended up holding seven of them there over the next few years. Sometimes he would feel playful and stack up a pile of RVs with the forklift, or unveil a surprising treasure from his indoor collection, just for us to shoot. He spontaneously cut a Lincoln Town Car in half with the T-Rex-sized demolition shears one sunset, as we all stood in a circle around The Pit, cheering and clapping. Always a showman, he loved that we loved the place.

Everyone who sees it loves it, but some love it too much. This part of the Mojave grows more desperate and poorer every year, and the locals rip him off constantly. Paul is full of stories: "Once, after being robbed, I asked a neighbor if she saw anything, because they cut a hole in the fence across from her and parked by it to load up. She recalled a red pickup there for a few hours the other night and heard them. I asked why she didn't call the cops and she said she was afraid they would come over and find her pot garden. So she sat on her porch smoking dope and watched me get robbed. Twice."

Paul has always had to work hard to stay ahead of vandals, thieves, and scrappers. He's hidden booby traps around the yard. Vehicles would be laid out in cramped mazes, making it difficult for thieves to carry things, and there's more razor wire atop the fence and gate every year. It's remote, but not remote enough. That's why we had every workshop attendee sign a non-disclosure agreement when it comes to revealing the exact location of Paul's. No, I will not say where it is, because Paul wants it that way.

Every trip is a little different. Things turn over fast at Paul's. There are fewer vehicles in the yard every year. It seems everything inexorably inches its way closer to The Pit, where they are consumed. Much of what you see here is already gone. Sadly, even the yard itself is facing extinction. As Paul says: "Now with California legislating to make all things diesel obsolete, I'm forced to retrofit or upgrade my already paid-for trucks and equipment. I'm not able, or willing to 'go buy new' as we're being told. That change of business is showing me that recycling has changed drastically. Used equipment or parts have less value daily. I'm in the process of cleaning up the yard to cash in on assets before they become a liability."

The remaining treasures at Paul's will find their way to his other yard, closer to town where he can keep an eye on them. That will allow this desolate spot to revert to the desert, like the other sites in this book. The desert always wins. It plays the long game.

Northro

Paul's Junkyard. 1/20/2011, 11:18 p.m.

Canon 20D, f/5.6, ISO 200. 1205 seconds, 3500K WB.

Full moon / white xenon flashlight and red-gelled strobe flash.

Twenty minutes of stars spinning around Polaris, behind a Northrop F-5 fighter jet tail section, damaged as a movie prop.

Swept Away
Paul's Junkyard. 1/19/2011, 11:43 p.m.
Canon 20D, f/5.6, ISO 200. 120 seconds, 3800K WB.
Full moon / white and red LED flashlight.

Tossed in a tidal wave of trash.

The Raptor Howls
Paul's Junkyard. 4/15/2011, 12:04 a.m.
Canon 60D, f/8, ISO 200. 2005 seconds, 5000 WB.
Full moon / red-gelled strobe flash.
This former sheriff's bus had a massive roll cage installed and was then barrel rolled by stunt driver Corey Eubanks for the movie *Fast Five*. It was eaten by that dinosaur within days of shooting this.

Boxy But Safe ►
Paul's Junkyard. 4/16/2011, 1:08 a.m.
Canon 60D, f/8, ISO 200. 167 seconds, 5000 WB.
Full moon / red and lime-gelled LED flashlight.
A *circa*-1990 Volvo Wagon hangs in the jaws of the demolition shears. The camera was directly under the car, pointed almost straight up.

SHERIFF'S

Iced
Paul's Junkyard. 4/4/2012, 9:12 p.m.
Canon 60D, f/8, ISO 200. 46 seconds, 5800 WB.
Deep shade / white LED flashlight and blue-gelled strobe flash.
A bullet-riddled postwar Chrysler from the 2013 movie *Gangster Squad*.

◀ **Pony Withers**
Paul's Junkyard. 10/9/2011, 12:57 a.m.
Canon 60D, f/8, ISO 200. 45 seconds, 5650 WB.
Deep shade / red and blue-gelled LED flashlight.
Longnose 1971 Ford Mustang plugging a hole in the north wall.

Tag the Sky
Paul's Junkyard. 10/10/2011, 7:25 p.m.
Canon 60D, f/10, ISO 200. 164 seconds, 5300 WB.
Full moon.

The last of the sunset over The Pit. There was very little color and lightness in the sky visible to the naked eye, but the time exposure allowed it to brighten and intensify. I love the idea of star trails appearing in a sunset.

2RWV538

Twin Cone Meat Wagon
Paul's Junkyard. 10/5/2014, 9:59 p.m.
Canon 60D, f/8, ISO 200. 281 seconds, 6050 WB.
Full moon / white, purple, and red Protomachines flashlight.

◀ **F+B Ambulance**
Paul's Junkyard. 5/13/2017, 11:07 p.m.
Canon 60D, f/8, ISO 200. 108 seconds, 3650 WB.
Total darkness / lime, purple, and blue Protomachines flashlight.
1970 Cadillac ambulance, complete and unrestored. It still had the gurney inside coated with a thick layer of dust.

AM in the PM
Paul's Junkyard. 1/21/2011, 1:06 a.m.
Canon 20D, f/5.6, ISO 200. 723 seconds, 5000K WB.
Full moon.
Parted-out monster truck on a shipping container, against twelve minutes of late-night sky.

Tri-Power Catfish ►
Paul's Junkyard. 9/27/2012, 9:49 p.m.
Canon 60D, f/8, ISO 200. 108 seconds, 3800 WB.
Full moon / white, red, blue, and green Protomachines flashlight.
Fifty-seven Chevy drag car, stored along the south fence.

The Wall in Space
Paul's Junkyard. 1/20/2011, 9:00 p.m.
Canon 20D, f/5.6, ISO 200. 723 seconds, 5000K WB.
Total darkness / white LED flashlight and strobe flash.

Film and TV prop cars, stacked and awaiting their fate in The Pit. Shot in total darkness, before moonrise.

The Purple Merge
Paul's Junkyard. 10/29/2012, 2:37 a.m.
Canon 60D, f/8, ISO 200. 229 seconds, 5700 WB.
Full moon / white, orange, and purple Protomachines flashlight.
Piled RVs and heavy equipment in The Pit.

C10 9/11 ►
Paul's Junkyard. 9/10/2011, 10:44 p.m.
Canon 60D, f/8, ISO 200. 238 seconds, 5000 WB.
Full moon / white, red, and orange-gelled LED flashlight.
Lightning strikes on the 10th anniversary of 9/11, when Paul raised the flag on the extended ladder of one of his old fire trucks. This pickup was cut in half for use as a movie prop.

ONE
WAY

Hereby Discharged
Paul's Junkyard. 9/10/2011, 12:07 a.m.
Canon 60D, f/8, ISO 200. 193 seconds, 5300 WB.
Full moon / red and blue-gelled strobe flash.
Summer monsoons and electrical storms are no stranger to this part of the Mojave.

Robot Forehead ►
Paul's Junkyard. 10/19/2013, 8:02 p.m.
Canon 60D, f/8, ISO 200. 76 seconds, 3800 WB.
Deep shade / white and red Protomachines flashlight.
A strangely upright, *circa*-1965, Kelsen Sports Rider. A cartoonish three-wheeled early electric car that was frighteningly street legal. Paul keeps this one—along with some other cherry rarities—stored in containers out of the elements.

Surreal Cruiser
Paul's Junkyard. 10/5/2014, 8:25 p.m.
Canon 60D, f/8, ISO 200. 206 seconds, 3800 WB.
Full moon / purple, lime, and red Protomachines flashlight.

Every Mile Magnificent ►
Paul's Junkyard. 10/5/2014, 11:53 p.m.
Canon 60D, f/11, ISO 200. 263 seconds, 3800 WB.
Full moon / white Protomachines flashlight.
A centerpiece of Paul's, the unrestored split-level GM Scenicruiser. Designed and built exclusively for Greyhound in the mid-fifties, only 1,001 were built. They suffered from driveline and metal fatigue issues. As iconic and fresh as they looked on America's new Interstates, most of these buses were scrapped by 1970.

The Split Level Craze ►
Paul's Junkyard. 1/5/2015, 11:07 p.m.
Canon 60D, f/8, ISO 200. 437 seconds, 3800 WB.
Full moon / blue and red Protomachines flashlight.

STEP DOWN
PLEASE WATCH STEP

NO BOARDING WHILE IN MOTION

Tramway to the Stars
Paul's Junkyard. 10/4/2014, 11:03 p.m.
Canon 60D, f/11, ISO 200. 3076 seconds, 3800 WB.
Full moon / purple, lime, and red Protomachines flashlight.

◀ **Junkyard Thrill Ride**
Paul's Junkyard. 10/19/2013, 1:03 a.m.
Canon 60D, f/11, ISO 200. 291 seconds, 3800 WB.
Full moon / white and red Protomachines flashlight.
The old parking lot trams for *Magic Mountain* amusement park passed through Paul's on their way to oblivion.

Mack's Pulsing Red Nose

Paul's Junkyard. 9/11/2011, 1:38 a.m.

Canon 60D, f/8, ISO 200. 253 seconds, 5100 WB.

Full moon / white, red, blue, and lime-gelled LED flashlight.

A late-thirties Mack fire truck along the south fence on a stormy summer night.

The Mummies
Paul's Junkyard. 3/17/2011, 12:20 a.m.
Canon 20D, f/5.6, ISO 100. 121 seconds, 5000 WB.
Full moon / white LED flashlight.
Nineties surfpunk band *The Mummies* 1963 Pontiac tour-ambulance.

Cop Crush
Paul's Junkyard. 10/5/2014, 10:51 p.m.
Canon 60D, f/8, ISO 200. 62 seconds, 3800 WB.
Full moon / white and red Protomachines flashlight.
A pile of wrecked movie prop police cars destined for the shredder.

Before the Feature Starts ►
Paul's Junkyard. 5/24/2013, 10:45 p.m.
Canon 60D, f/8, ISO 200. 2527 seconds, 3850 WB.
Full moon / purple and red Protomachines flashlight.
One of two *Cobb* movie prop 1959 Imperials bizarrely perched on gangly legs with tiny casters.

Blue-Eyed Soul
Paul's Junkyard. 4/23/2013, 8:42 p.m.
Canon 60D, f/8, ISO 200. 178 seconds, 3800 WB.
Full moon / white, orange, and blue Protomachines flashlight.
The relentlessly squaresville 1971 Ford LTD. The sedan of choice for every seventies cop show.

The Other Split Window ▶
Paul's Junkyard. 10/27/2012, 11:09 p.m.
Canon 60D, f/8, ISO 200. 43 seconds, 3900 WB.
Full moon / red and lime Protomachines flashlight.
The *Lemony Snicket* Tatra 603 coated with Mojave dust.

KENWORTH

◄ **Blue Radius**
Paul's Junkyard. 1/19/2011, 11:07 p.m.
Canon 20D, f/5.6, ISO 200. 474 seconds, 5000K WB.
Full moon / white and lime-gelled LED flashlight.
Kenworth, looming.

The Graveyard of Southern Rock
Paul's Junkyard. 3/16/2011, 8:12 p.m.
Canon 20D, f/5.6, ISO 100. 31 seconds, 5600K WB.
Total darkness / white and blue-gelled LED flashlight.
A case of 8-track tapes, rotten and decaying inside the collapsing workshop. These were mostly Skynyrd and Allman Brothers albums. And one by Uriah Heep.

Howler de Ville
Paul's Junkyard. 4/27/2018, 10:51 p.m.
Canon 60D, f/8, ISO 200. 108 seconds, 3900 WB.
Full moon / white and red Protomachines flashlight.
The black seventy-two Caddy by the workshop, the one that hasn't moved in over twenty years, on a night of ceaseless, make-your-tripod-fly-away wind.

Watch Your Fingers ►
Paul's Junkyard. 5/25/2013, 8:02 p.m.
Canon 60D, f/8, ISO 200. 467 seconds, 5000 WB.
Total darkness / white and red Protomachines flashlight.
Semi-truck chassis—just an engine, gas tanks and exhaust stacks—sitting in the sand. The wind was slowly spinning the fan on the front of the engine, so I spun it faster with my hand and lit it with a flashlight. Shot in total darkness before moonrise.

◀ **The Monster in the Backseat**
Paul's Junkyard. 9/13/2011, 12:28 a.m.
Canon 60D, f/10, ISO 200. 61 seconds, 7000 WB.
Full moon / red and lime-gelled LED flashlight.

Snicket Float
Paul's Junkyard. 5/26/2013, 12:09 a.m.
Canon 60D, f/8, ISO 200. 96 seconds, 5550 WB.
Full moon/ blue and red Protomachines flashlight.

Count Olaf's Eyes
Paul's Junkyard. 9/12/2011, 9:05 p.m.
Canon 60D, f/10, ISO 200. 109 seconds, 3800 WB.
Full moon, dark interior / white, red, and lime-gelled LED flashlight.
1960 Chrysler Imperial limo, modified for use as a prop in the 2004 *Lemony Snicket* movie. The custom dashboard is unlike anything I've ever seen in a junkyard.

US NAVY
91-04207

◀ **The Wayward Bus**
Paul's Junkyard. 4/23/2013, 11:21 p.m.
Canon 60D, f/8, ISO 200. 467 seconds, 3800 WB.
Full moon / white, red, and lime Protomachines flashlight.
Navy bus perched on a pile of debris. It was rendered in The Pit days after this shoot.

Space Station Number 8322
Paul's Junkyard. 5/26/2013, 10:41 p.m.
Canon 60D, f/8, ISO 200. 243 seconds, 5000 WB.
Full moon/ red Protomachines flashlight.
NYC subway cars in space.

VCR Repair
Paul's Junkyard. 4/8/2012, 1:00 a.m.
Canon 60D, f/8, ISO 200. 122 seconds, 3300 WB.
Full moon / white and red LED flashlight.
There were three of these identically painted late-seventies Dodges in the yard, with various levels of car chase damage.

A Short Hop Into Hell ►
Pauls Junkyard. 10/20/2013, 10:07 p.m.
Canon 60D, f/8, ISO 200. 142 seconds, 3850 WB.
Total darkness / lime, purple, and red Protomachines flashlight.
A difficult set up in the grimy cockpit of a Bell Huey helicopter wreck. The broken floor was patched with sheet metal, so the camera wobbled if you were inside moving around. It forced me to do all the lighting from outside.

Kingfish
Paul's Junkyard. 1/20/2011, 1:15 a.m.
Canon 20D, f/5.6, ISO 200. 482 seconds, 3900K WB.
Full moon / white xenon flashlight and red-gelled strobe flash.
One of the ex-military Bell Huey UH1 Iroquois helicopters decaying on the lot.

Time Tunnel ▶
Paul's Junkyard. 4/7/2012, 11:37 p.m.
Canon 60D, f/8, ISO 200. 146 seconds, 3600 WB.
Total darkness / white, red, blue, and green Protomachines flashlight.
Looking into the tail of a Vought F8 Crusader II with its engine removed.

Rocket New Yorker

Paul's Junkyard. 8/15/2016, 10:46 p.m.

Canon 60D, f/8, ISO 200. 123 seconds, 3800 WB.

Total darkness / red and blue Protomachines flashlight.

Seventy-one Chrysler, modified for a superhero movie. Paul tells me it never appeared on screen in the finished movie, but it disappeared in the shredder a few days after shooting this.

KLENK
SECURITY

Heist
Paul's Junkyard. 1/20/2011, 12:02 a.m.
Canon 20D, f/5.6, ISO 200. 120 seconds, 5300K WB.
Full moon / white LED flashlight.
A steady stream of shot up and crashed prop armored cars have passed through Paul's. I photographed the Klenk truck a half hour after tearing my rotator cuff in a fall.

◀ **Klenk**
Paul's Junkyard. 7/13/2019, 11:24 p.m.
Canon 6D Mk II, f/8, ISO 200. 240 seconds, 3900 WB.
Full moon / white Protomachines flashlight.

CITY of L.A.

Flintstones Xtra
Paul's Junkyard. 4/27/2018, 10:28 p.m.
Canon 60D, f/8, ISO 200. 123 seconds, 4150 WB.
Full moon.
Made of foam, this crumbling sedan from one of the nineties *Flintstones* movies was never meant to survive beyond filming.

◄ **City of LA**
Paul's Junkyard. 5/13/2017, 12:58 a.m.
Canon 60D, f/8, ISO 200. 180 seconds, 3800 WB.
Deep shade / white Protomachines flashlight.
Cabover garbage truck, now garbage itself. The cab tilted forward, with the roof cut off and no driver's seat.

4

TECHNIQUE

The photographs in this book were shot at night. The full moon serves as the primary light source in most of these images. Some were shot in the total darkness of deep shade. Many are a mixture of both.

The basic exposure formula: ISO 100 or 200, at f/5.6 or f/8, with most exposure lengths running from one to four minutes. All the images were captured with Canon 20D, 5D MkIII, 60D and 6D MkII DSLRs, and a variety of wide-angle lenses, ranging from 30mm to a fisheye.

The color and light effects were all done in-camera during the exposure. I don't use light stands, umbrellas, synced lighting or external power supplies. All my lighting is hand-held and added one source at a time during the minutes-long exposures. I normally light the entire scene with a single light source, used from multiple angles.

For the 2006 through 2011 images, I used Maglite and Coast brand xenon and LED flashlights and a single Canon 430EX strobe—all frequently masked with theatrical lighting colored-gel material.

The post-2012 work was lit with the Protomachine LED, a 100% hue-, saturation-, and brightness-controllable flashlight that fits in your coat pocket. Literally millions of colors, at any brightness, at the touch of a button. It's bright enough to light a stacked pile of cars from a hundred yards away, but adjustable enough to subtly light a screw head from two feet.

I also use hand-held reflectors, snoots and gobos to manipulate the way the light falls. This simple lighting methodology allows me to travel light and work fast in these difficult, often cramped locations.

These photographs were finished in the digital darkroom. Some are composites, or star trail stacks, using multiple exposures from the same tripod setup. They may also incorporate more traditional, yet still digital adjustments, such as exposure, contrast, and spotting. In all cases, however, the lighting and color were done in-camera. These images are not Photoshop creations. What you see is what I saw and lit that night.

Carbecue
Milligan, California. November 1990.
Canon FX, f/5.6, Kodak 160T Chrome film. 8 minutes.
Full moon / red and green-gelled strobe flash.
1961 Chrysler, mysteriously placed on top of a concrete foundation and shot to death, at a now forgotten mining ghost town, way down a dirt road in the SoCal desert.

Growing up, I was a sixties California car-crazy kid, obsessed with Hot Wheels, Ed "Big Daddy" Roth, and *Car-Toons* comics. I grew up to become a *Micro Machines* toy car designer at Galoob Toys in the nineties and a freelance artist after the turn of the century. The aesthetics of cars and car culture occupy a big room inside my head. My UrbEx preoccupation was triggered by teenaged trips to Old West ghost towns, sneaking around abandoned World War II bunkers, and trolling recently bypassed highway towns in the late seventies. Given this background, it's only natural that I would gravitate toward shooting abandoned cars, right from the start.

I started creating hand-lit pictures in the dark in 1989 after auditing a night photography class taught by Steve Harper at the Academy of Art in San Francisco, where my brother Tom was enrolled. I was floored by the moonlit, time-exposure aesthetic, immediately connecting it to the creepy abandoned sites I knew from my past. Having no previous experience with manual photography, I bought a used 35mm camera and made my first baby-steps under ridiculously complex circumstances: experimenting with eight-minute exposures by moonlight of abandoned Route 66 buildings.

Over the decades I've brought this style of endless trial and error, evolution and play, to capture a never-ending stream of decommissioned military bases, industrial infrastructure, junkyards of every description, bypassed and abandoned roadside towns, aircraft boneyards, derelict amusement parks, hospitals, hotels, and ocean liners. The *Lost America* body of work is a thirty-year—and still ongoing—art experiment.

Table of Contents

What Do I Do Now?

You weren't expecting it. You didn't see it coming, and it caught you completely by surprise. Life has taken a sudden, drastic turn, and a new reality has taken its place. Unfortunately, in this reality, your very survival is at stake.

You're hungry, thirsty, tired, cold, and suddenly alone. All the resources that were easily accessible to you just a short time ago have suddenly disappeared. All the government systems and agencies that you count on regularly are unavailable. You can't be certain what's really going on because your ability to communicate with others is limited. Cell phones? Not now. There's no water coming from your faucet. There's no power in your home, either, so refrigeration is not an option. What about your food? What if your cupboards are empty and there's nothing available to purchase at the stores? You *must* eat to survive, and you *must* have water. In fact, water is more important than almost anything else. But where will you get these precious things that your well-being—and your very life—depend on? What will you do, and more importantly, how will you survive?

Or *will* you?

THINK LINK

- What types of things could happen to create a **doomsday scenario**?
- What basic skills do you think are necessary to help you survive?
- Describe any **survival skills** you already have.

What *Is* Survival?

Survival has become a popular topic in recent years. There are television shows that depict survival situations. And there are reality game shows in which survival is part of the game. You can even learn about survival in books, Internet videos, blogs, websites, and magazines.

But what is survival exactly? It depends on whom you ask. You are likely to find many different answers.

At its root, *survival* means to keep living. No one lives forever. But most people want to live long, enjoyable lives. Much of modern life has been dedicated to extending the human life span. At the same time, people have become dependent on **industry** to support survival. And we have created lifestyles that may separate us from more traditional, and perhaps even natural, ways to survive. Most people in modern cultures lack the skills that humans have depended on to survive and thrive for thousands of years.

I Didn't Expect That!

What could possibly happen to create a doomsday scenario? The truth is, such a scenario is highly unlikely. The following types of events that require survival skills are more likely to happen:

- an accident in the wilderness far from help (a rock slide, a tumble down a steep slope, etc.)
- a natural disaster, such as an earthquake, a tornado, or a hurricane
- getting lost in the wilderness
- an automobile accident or breakdown in the middle of nowhere

Methods and Tools

Most of us depend on our modern way of life to survive. Healthy, clean, hot and cold water are available in our houses at the turn of a knob. We have heaters to keep us warm. Air conditioners keep us cool. We can buy clothes easily in stores. Food is available in restaurants and stores all around us. So, what can we use to help us learn to survive in emergency situations?

Today, learning to survive involves the use of methods and tools. What methods and tools are needed? Well, some methods may involve having different survival **strategies** at your fingertips. The strategies can be used in any number of unusual situations. For example, one strategy may be ways to find clean water. Another may be methods to keep your body warm. The tools are devices people can use as part of each strategy to survive. They may be things we already have, such as knives, blankets, or other things made in factories. They may also be **primitive**, or handmade from **repurposed** or natural materials. One survival strategy may be knowing how to make these things!

Changes to a person's current way of life may have **dire** consequences. That is especially true if the person doesn't have the basic knowledge needed to survive.

You Need Every Drop

Experts agree that you need about two quarts (64 ounces) of drinking water each day. In extreme weather or if you exercise a lot, you need about a gallon (128 ounces) a day. Under normal conditions, how much water would a family of four need to survive one week, in quarts, ounces, and gallons?

Everyday Carry

To be prepared for emergency situations, some people carry small basic tools with them all the time. These tools might include a whistle or a multi-tool pocket knife. These are called everyday carry, or EDC. Equally smart is learning to make tools out of objects found in nature.

The Most Important Things to Remember

There is a lot to learn about survival, and it can feel overwhelming. Most experts agree, though, that there are a few key principles to remember. These key principles will help increase your chances of survival.

Preparation

The longtime scouting motto is "Be Prepared," and there's a good reason for that. It is important to be aware of your location and home environment. When you travel, it is important to be aware of your destination's environment, too.

Another aspect of **preparation** involves being properly skilled. It is important to prepare yourself in case of emergency situations. Do you know how to start a campfire and keep it going? Do you know how to make a shelter? Do you know where and how to find water and food? And if you do have these skills, do you also have the proper equipment and supplies available?

Deep Breaths

If you find yourself becoming overwhelmed, sit down and breathe. Taking long, deep breaths calms the body and mind and gives you time to return to more relaxed levels.

Fight or Flight?

Maybe you do begin to panic. You think dire thoughts, your heart rate increases, you feel short of breath, and your body starts to shake. You are primed to fight—or ready to run. This is called the *fight or flight response*. It is a natural response that has helped humans and other animals survive challenging situations.

Composure

In a survival scenario, nothing matters more than **composure**. Survival expert Cody Lundin outlines a fivc-point plan to keep anyone in the right frame of mind. He calls it ***STOPA***: **S**top, **T**hink, **O**bserve, **P**lan, and **A**ct. Following the five steps will help you maintain your composure.

In emergencies, we can become fearful, but panicking makes situations worse. Be aware of your emotions, and try your best to manage your responses. Especially in an emergency, this will enhance your chances for survival.

Strategy

It's important to come up with a plan to help keep you focused. Develop a strategy. You can do this by observing and assessing your environment. Then, think about your **priorities**. What things do you need to consider first? The following are key priorities to think about:

- secure personal safety
- maintain body temperature
- maintain body **hydration**

Putting these three things first, preferably in this order, will go a long way toward your survival. Consider these, and then decide on a response to your situation.

In the Zone

Normal average body temperature is 98.6°Fahrenheit (37°Celsius). It is normal for a person's temperature to go up or down from that number by about 1.1°F (0.6°C). What is the normal range of body temperature in both Fahrenheit and Celsius?

Action

You've kept your cool, followed priorities, and made a plan. Now, it's time to act on your decisions and commit to them. But stay aware, and keep assessing. Continually assess your environment, circumstances, and personal conditions. How is your plan working? Are you getting the desired results? Remember to **adapt** as needed. This will increase your chances of survival.

If your plan includes moving from one place to another, you might discover a new option. Be ready to take advantage of any new opportunities and circumstances. Maybe you planned to build a shelter, but in looking for materials, you find an existing shelter. Maybe you find an unexpected water source. Or perhaps you come upon animals or people that may be dangerous. In any case, be smart and be patient with yourself and your decisions.

THINK LINK

In a survival scenario, it's important to think and get oriented in terms of time and location. Ask yourself the following questions.

- **Time:** What time of day and what time of year is it? How much daylight do I have left? About how long will darkness last? Is air temperature going to get hotter or colder?
- **Location:** From which direction did I come? Which direction is uphill? Which direction is downhill? How far am I from the nearest community? Where is the nearest emergency station (firehouse, police station, etc.)?

DIG DEEPER!

Can You Survive?

The most important tool in a survival scenario is your brain. You've got to think about and assess your current conditions. After a little time passes, think again. Conditions change, and you should always adapt your plans to the conditions.

Do you have what it takes to survive? Can you successfully assess your situation? Take this quiz to find out!

1. *You are lost in the wilderness. You . . .*
   a. panic.
   b. climb a tree.
   c. hunt for food.
   d. think about the resources you have and how to use them.

2. *Your car has broken down in the middle of the desert. You . . .*
   a. go to sleep until someone finds you.
   b. drink all your water.
   c. start walking because you're bound to run into help.
   d. think about the conditions you are in and make a plan to survive them until you can change them.

3. *You've just survived a terrible tornado. Everything around you is flattened. You . . .*

   a. just sit and cry because, after all, you're in a tough situation.

   b. randomly start poking in the wreckage.

   c. hide because another tornado may be on its way.

   d. look around, gather resources, and make a plan for survival.

If you answered *d* to every question, you're on the right track! Always use your brain to assess . . .

- yourself
- the situation
- the environment
- your tools and equipment

Remember: You don't have to guess! You can be your best and pass any test if you simply assess!

Water Is Life and Death

"When the well's dry, we know the worth of water."
—Benjamin Franklin

You know and understand now the basic keys to survival. But none of that knowledge means much of anything without water. Water is essential for life. For people, it is even more important than food. This is because water is a big portion of our blood and is in all our cells. We can't live without it.

The specific amount of water each person needs depends on his or her size and age. It also depends on the outside temperature and the body's activity. For example, the more you sweat, the more water you need to **replenish** what is lost.

People lose water in different ways. It is lost mainly through urination, perspiration, and respiration. These are part of the way the human body works to keep us healthy, so there is no way to avoid them. But perspiration and respiration can be limited to help maintain hydration.

Amazing Camels

Some camels can survive about 180 days without drinking water. The longest humans can survive without water is about 7 days. How many times longer can camels survive without water than humans?

Did You Know . . .

. . . you can become dehydrated in freezing weather? This is because there isn't much moisture in cold air. In extreme cases, simply breathing can dehydrate you.

Changing Attitudes

The first peoples camped or lived near steady and reliable water sources. To stay hydrated and healthy, civilizations were developed near these bodies of water.

Today, in many parts of the world, people tend to take water for granted. We can get water so easily that there isn't much reason to think about it. Unfortunately, easy access to water can make us treat it without care. We throw trash and debris into oceans, lakes, rivers, and streams.

Early people did not treat water so carelessly. They had great respect for water. They knew how vital it was for life. It seems the more civilized we became as societies, the more poorly we treated our water supplies. Things like water filters and bottled water made it easier for us to mistreat natural sources.

Beware Bacteria!

Bacteria in **tainted** water may make a person sick. At best, water bacteria can cause diarrhea, vomiting, and stomach cramps. At worst, bacteria can eat away at the body and cause the thirsty person to become gravely ill and eventually die. So, be careful before you start drinking any untreated water!

There is bacteria in this water.

When Water Is Needed

We are used to having drinking water available. But in a survival scenario, water may not be easy to get. It may have to be found. And if that's the case, you must keep a few things in mind:

- **Quality:** Is the available water polluted or **unpotable**?
- **Access:** Is the available water in a dangerous or difficult-to-access location?
- **Equipment:** Do you have what you need to boil the water for safe drinking?

Safety First!

If you don't know with absolute certainty that water is safe to drink, boil it or treat it first . . . *always*! See pages 26–27 for more information about how to treat water.

Cities are responsible for protecting their water sources.

Conserving Bodily Fluids

In an emergency, you need to conserve the water that is already in your body. But how do you restrict water loss?

During the day, get into the shade and stay there. It also helps to minimize physical activity when it is hot. If you must be active, do it in the evening when the sun has gone down but it is not fully dark yet. You can also be more active in the early morning when there is enough light to see but the sun has not fully risen. And be sure to perform any activities at steady, even paces to minimize overexertion and perspiration.

Digestion uses a large amount of water in the body. Remember, you can live much longer without food than without water. So, unless you absolutely need to, do not eat.

Even if you're hot, keep your clothes on and cover your head as well. You will lose more moisture if your skin is directly exposed to the air. Try to stay out of the wind as much as possible, since wind also increases water loss.

If you have an ample water supply, make sure you drink lots of clean water. Don't take even a small chance of dehydrating yourself.

The Pinch Test

To test a person for dehydration, pinch the skin on the back of his or her hand. If the skin snaps back quickly, then the person is likely fine. If the skin returns to its normal shape slowly, the person may be dehydrated.

From an Expert

Survivalist Dave Arama offers this advice for conserving water and energy in our bodies:

- If you don't have to stand, sit.
- If you don't have to sit, lie down.
- Minimize talking, which takes effort and increases respiration.
- Breathe through your nose, since mouth breathing releases more water.

Finding Water

Remember the survival priorities mentioned earlier? If you do, you know what you need to do first. You need to secure your safety and develop a plan to address **exposure**. Assuming you have done those things, you can and should address your body's need for a steady supply of water. You should start to think about and plan how to find water.

The key to finding water is not necessarily the method to collect it. Instead, the key is understanding water itself and knowing where to find it. In other words, it's not *how* to get it so much as *where* to get it.

Tall Drink of Water

Adults are made up of about 60% water. The average adult human body has a volume of 66.4L. So, how much of that overall volume is water?

Movies often show heroes in dire straits, using clever methods to obtain water. But most movies' goals are to create drama and suspense, not to depict reality. Remember that these movies are not instructional videos. Do not count on the film industry to provide practical guidelines for survival. The methods may look cool, but if you try to follow them to protect your life, you're likely to wind up very thirsty—or worse.

So, how do *you*—a real person in the real world—find water?

Heat Exhaustion

Prolonged exposure to heat can result in heat exhaustion—or worse, heatstroke. Signs of heat exhaustion include heavy sweating, dizziness, weak and rapid pulse, feeling faint, nausea, headache, and more. Heatstroke may include all these plus fainting, a high fever, and hot, dry skin because the person has stopped sweating from loss of water.

First things first. Look around you to check out water availability in obvious sources such as creeks, streams, and lakes. You can also look for snow, ice, or slush, which are merely frozen water. While looking around, you'll quickly be able to answer, "Yes, I have a water supply" or "No, there's no water in sight."

Yes, I Have a Water Supply

If your answer is *yes*, then you are in relatively good shape. You just need to make sure that the water is clean enough to drink. You should assume that all water is **contaminated**. You'll need to treat any and all water before you drink it, unless you are completely certain the water is safe. (Read about treating water on pages 26–27.)

If snow, ice, and slush are available, they can be melted to drink, but you'll need a container to store the water. Of course, just as with any water, be sure to decontaminate it by boiling or treating it so it's safe.

Signs of Dehydration

Dehydration is a serious issue. It is important to know the symptoms. Symptoms of dehydration include headaches, nausea, poor judgment, depression, and dark yellow urine. If a person is not urinating at all, then he or she is likely dehydrated.

No, There's No Water in Sight

If your answer is *no*, you will need to find water in a situation where it is not obvious. There are many ways to do this, and some will surprise you. Keep reading to find out how to do this!

DIG DEEPER!

Treating Water

Having safe water to drink is extremely vital. So, knowing how to treat water is an essential skill to learn. Here are three ways to treat water so that it is safe for drinking.

Boiling

This is the easiest method to decontaminate water. Boil the water for 10 minutes. In high elevations (above 5,000 feet or 1,524 meters), boil it an additional minute for each 1,000 feet (304.8 meters) of elevation above sea level. Keep the pot covered to avoid as much evaporation as possible. Be sure to let the water cool before drinking it.

How Long?

Determine how long you will boil water to purify it at 5,000 feet. How long at 7,000 feet? How long at 10,000 feet?

Bleaching

Using an eyedropper, add 16 drops of chlorine bleach to each gallon (3.8 liters) of water. Use liquid chlorine bleach that is 5% to 6% sodium hypochlorite with no additives or preservatives. Mix and let stand for 30 minutes. The water should smell slightly of bleach. If it doesn't, add 16 more drops and let it stand for 15 minutes. If it still doesn't smell of bleach, the water is too contaminated to drink.

Distilling

This method is a lot harder and less successful than the other two, but it works with the right equipment. With this method, you can also rid saltwater of salt to make it drinkable. Put the water in a large pot. Tie a cup to the pot handle, with the opening of the cup placed toward the pot lid. Place the lid on the pot. The cup should not touch the water. Boil the water for 20 minutes. Evaporating water that is clean and distilled will hit the lid and condense into the cup. There won't be much of it, but it is safe to drink.

Observe the Environment

One crucial way to find water is to study the **topography**. There is water in most soils; however, finding enough water to drink is the challenge. Sometimes, finding water can be as easy as walking. Walking downhill might lead you to a water source. Water flows downhill and collects in valley bottoms. Even the water flowing underground flows downhill. Seeps and springs will probably be located a little farther up than the valley bottom, but they can also be found at many locations along the valley—so pay close attention. Creeks and streams may flow back underground before running along the surface. There may also be water in what appears to be dry riverbeds or creek beds, especially at their bends, but you may need to dig down to find it.

Water may be in **catchments** under rocks as well. And pay special attention to the color and location of vegetation. Fertile and healthy vegetation must have a water source to grow, so look around and dig.

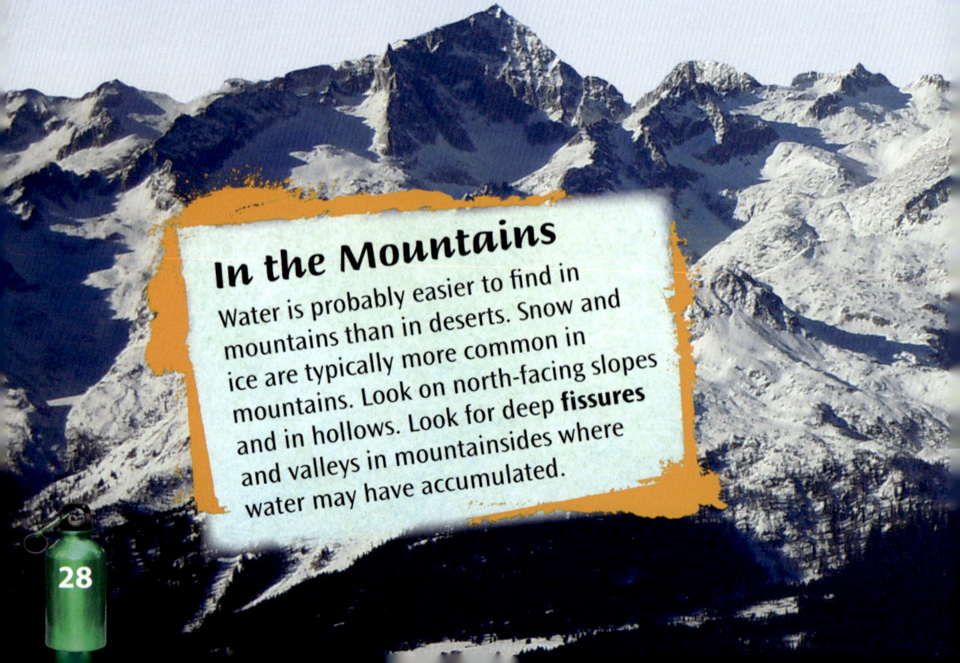

In the Mountains

Water is probably easier to find in mountains than in deserts. Snow and ice are typically more common in mountains. Look on north-facing slopes and in hollows. Look for deep **fissures** and valleys in mountainsides where water may have accumulated.

Follow Animal Trails

Animals aren't just fun to watch—they can also help you find water. Birds, insects, and small animals that live in an area know where the water is. They couldn't live there if they didn't. Follow them, and you may find water.

In the Desert

Prickly pear cactus is a plant found in many deserts. To get the water from the plant, scrape the needles from the young prickly pear pads, cut the pads into pieces, and chew them. You don't really extract the water, but you "eat" the water instead.

prickly pear cactus

Collect Dew and Rain

In many places, there is condensation in the form of dew in the early morning hours. Dew can be collected from the leaves of some plants, or it can be gathered on a tarp or sheet of plastic. Ideally, the dew can be collected in a container to store and use over time.

Know What You've Got

You have a 55-gallon rain barrel that is 60% full. With the amount of water currently in it, how many gallons of water are available to you in an emergency situation?

Seawater

It shouldn't be a problem to get safe water from the ocean, right? Wrong. Trying to consume salty seawater can be challenging or even impossible. If it rains, take advantage of the great opportunity. Catching rainwater will help you make salt water safer to drink. Rainwater can be mixed with seawater to increase the supply. To be safe, make sure you mix in at least half rainwater. In a survival situation, drinking a mixture of sea water and rainwater is all right.

Rain can be a saving grace in a survival situation. You can drink rain as it falls, or you can collect it in a container. Collecting rainwater is a great way to ensure you have a water supply to last you. If you don't have a container, you can make one using a sheet of plastic. You can dig a hole and line it with the plastic to catch rainwater. Shells, bark, and other natural materials, if shaped in the right way, may also be able to catch and hold rainwater. The more containers you have, the better.

Make a Vegetation Still or a Solar Still

What if you have done some searching and some digging and have still not found any water? Well, you still have options. The following two water-collection ideas are a little more involved but have been shown to work. One option is a vegetation still, and the other is a solar still. For these, you need a couple of key items.

For the vegetation still, you need a large, clear plastic bag and some cord. Plants release water through their leaves back into the atmosphere in a process called **transpiration**. To capture the water, tie a plastic bag around the leafy end of a tree branch or bush. The more leaves, the better. Willow or oak trees are probably best, but others work also. Tie the cord tightly around the opening of the bag so no air can escape. Wait for a while, but check the bag occasionally to make sure the cord is tight enough and the still is working properly. After a while, water released from the leaves will start collecting in the bag. When there is enough to drink, carefully remove the bag from the plant so that you don't spill the water or tear the bag.

Success!

Survival expert Christopher Nyerges has performed nearly 100 experiments on the two stills. He has found that vegetation stills produce 25% to 50% more water than solar stills.

Beware!

Don't use poisonous plants such as oleander, poison oak, and poison ivy for either of the stills. The water produced from these plants will cause intense physical reactions or even death.

vegetation still

Do the Math

If a solar still produces 1 liter of water over a day, about how much more water might you expect from a vegetation still of the same capacity over 7 days if it produces 25% more water? How about 50% more water?

To make a solar still, you need a large sheet of plastic and a wide-mouth container. You will be collecting water that evaporates from the ground. Find a location where you suspect the soil underground may be moist. Dig a hole about three feet (one meter) wide and three feet (one meter) deep. Place the container in the middle of the hole and surround it with plant material, such as grass or cactus. Place the plastic sheet over the hole, and pull the edges so it is suspended over the hole. Place rocks on the edges of the plastic to hold it in place and seal the hole. Place a small stone in the middle of the plastic, directly over the container. This should cause the middle of the sheet to drop below the level of the edges.

During the day, the sun will cause water to evaporate from the soil and plant material. Water will collect on the underside of the plastic, flow downhill, and drip into the container. Periodically check on the still to see how it is working and whether adjustments are needed.

Natural Disaster

In natural disasters like earthquakes, your water supply may be cut off. However, your house may still have plenty of water. Water heaters can hold about 189 liters (50 gallons). Water can even be found in pipes or in toilet tanks. And, of course, swimming pools hold several thousand liters of water. But be careful! Swimming pools usually contain chlorine. This water may need to be decontaminated.

Believe It or Not

It may sound gross, but you can urinate into the hole for a solar still. The water in urine will evaporate just like other water. And distilled (evaporated) water is quite pure and safe to drink.

In Consideration of Food

In survival situations, the first thing that many people think about is having enough to eat. This makes sense for many reasons. We all need food to live. Food gives the body fuel, structure, and **nutrients**. For many of us, our daily schedule revolves around meals. But in a survival situation, food is not always a priority. Most of us can live adequately without food for about three weeks. Looking for, preparing, and processing food can use up important time and energy. That time and energy might be better spent building a shelter or finding water.

Emotional Boost

Sometimes in normal life, people turn to food for comfort. That's not always a healthy choice. But in a survival situation, food may also provide emotional comfort and a psychological boost. It is important to remember, though, that survival depends more on water than food.

Side Effects of Fasting

While we can go quite long without food, it's not an easy thing to endure. There are hunger pangs, and you might notice sluggishness, fatigue, or loss of energy. The best thing to do is be aware of what you are experiencing, get some rest, and conserve your energy.

Also, after two or three days of **fasting**, the body's metabolism begins to change. After two weeks, the body uses the fuel and nutrients already stored in it much more efficiently.

Finally, digestion requires a lot of water. So, if you have a very limited water supply, it's probably not a good idea to eat. Only eat if you have a reliable and stable supply of water on hand.

WARNING

Many wild plants are poisonous. Never eat any wild foods unless you know for sure that they are safe.

Learn Your Local Foods

When it comes to food, nothing is more important than planning and preparing. If you're in a city or a town during a survival situation, finding food is probably less difficult. But what do you do if you are in the middle of nowhere, surrounded only by nature and wilderness?

Getting food from the natural environment is not an easy task. When it comes to **foraging**, hunting, and trapping, there are no shortcuts. This is where planning and preparing come in. To obtain food, you must know what sorts of plants and animals are available. It is important to study and develop skills if you want to have food in a survival situation. For instance, it can be helpful to know which plants are safe to eat. Developing these skills takes time and effort! You can read books to help you, or you can spend time in nature exploring with experts.

Any time and effort spent learning about local foods will also build confidence and awareness. It is rewarding and provides more than a little peace of mind. You'll know what to look for and what to do with it once you find it!

Acorns

Acorns are the fruit of oak trees, and oak trees are found all over the world. To eat an acorn, first crack the shell. Remove the seed, or meat, from the shell. Then **leach** the seed. You do this by repeatedly boiling the meat in water for one to three hours to remove the bitter taste. You will need to change the water several times while doing this.

Prickly Pear Cactus

You already know that a good source of water in the desert is the prickly pear cactus. The pads can also be eaten raw or cooked. Prickly pears produce fruit that can be eaten as well, and the seeds inside them can be ground and eaten.

Natural Disasters

Unfortunately, different kinds of natural disasters, including earthquakes, tornadoes, and hurricanes, happen regularly in various places. Many people may lose their lives in these events, and even more people may be left without the resources they need for survival. Such resources include water, food, candles, and blankets, among other things. What can people do to get the food they need in these disruptive situations?

Insects . . . Yum?

Insects are a healthy and nutritious source of nutrition. Many contain as much protein per pound as meat or fish. It is recommended that adult men should take in about 60 grams of protein each day. A 3.5-ounce serving of raw grasshoppers contains about 15 grams of protein. How many ounces of grasshoppers would an adult man need to eat to take in the recommended 60 grams of protein in one day?

The best thing to do is to be ready. We know that these events will happen again sometime. We don't know exactly when or how things will turn out, but we do know they are coming, so it's important to plan and prepare. One way to plan ahead for such disasters is by stocking up on **nonperishable** food items. Store dry foods, canned foods, and dried foods. Dry foods include nuts, seeds, and grains. Canned foods include canned corn, peas, and green beans. And dried foods are things such as dried fruit, beans, rice, or jerkies. Foods prepared in these ways can be stored for long periods of time and remain **edible**. Be sure to store nutritious foods you like to eat.

Be Prepared and Keep Your Cool

Even in a dire situation, there's always hope as long as you remain calm and use your brain. And most importantly, things are guaranteed to be even better if you stay thoughtful and prepared. That doesn't mean you always have to worry about bad things happening and try to prepare for every possibility. That's impossible, and more than that, worry takes the joy out of life. When we look at things realistically, we realize that most days, everything is fine. Survival isn't typically an issue, but challenges do arise. One day, you may need to rely on your skills, preparation, and brainpower to get through an emergency. If so, don't panic! You are up to the task. In fact, you can help others by staying calm, keeping cool, and remembering your survival priorities:

- secure personal safety
- maintain body temperature
- maintain body hydration

You ***know*** what to do. Now, get out there and do more than just survive—*live*!

Those Amazing, Adaptable Human Beings!

Many factors have helped the human species survive for more than 200,000 years. One of the most important is the fascinating and **analytic** human brain. Other factors include adaptable bodies and the ability to cooperate with one another. You have these same capabilities. So use them to their fullest!

Glossary

adapt—to change to fit a new or specific situation or circumstance

analytic—thinking about something by focusing on its parts

catchments—things that catch water

composure—calmness of mind

contaminated—infected, soiled, or tainted

digestion—the process of breaking down food inside the body to get its nourishment

dire—horrible; disastrous

doomsday scenario—a situation that marks the end of the world as it has been known

edible—safe to be eaten

exposure—the act of being open to extreme weather conditions for a dangerous amount of time; unprotected

fasting—not eating for periods of time

fissures—small openings or cracks

foraging—the act of looking for and gathering food from the wild

hydration—the act of having a sufficient supply of fluids, especially water

industry—business and manufacturing

leach—to pass a liquid through something to carry away unwanted and dissolvable parts

nonperishable—unlikely to spoil or decay

nutrients—things that provide the substances necessary for a body's health and growth

preparation—the action or process of making yourself ready in advance

primitive—quite basic; made by a self-taught artisan

priorities—things that deserve attention before others

replenish—to fill up again

repurposed—recycled; remade from existing materials

strategies—plans

survival skills—abilities to provide for the basic needs of life

tainted—contaminated; spoiled

topography—the shapes, heights, and depths of the land

transpiration—the process by which plants give off water vapor

unpotable—unsuitable for drinking

Index

Check It Out!

Books

George, Jean Craighead. 2004. *My Side of the Mountain*. Puffin Books.

Nyerges, Christopher. 2014. *How to Survive Anywhere: A Guide for Urban, Suburban, Rural, and Wilderness Environments*. Stackpole Books.

O'Dell, Scott. 2010. *Island of the Blue Dolphins*. HMH Books for Young Readers.

Olsen, Larry Dean. 1990. *Outdoor Survival Skills*. Chicago Review Press.

Paulsen, Gary. 2006. *Hatchet*. Simon & Schuster Books for Young Readers.

Rice, William B. 2013. *Survival! Desert*. Teacher Created Materials.

_____. 2013. *Survival! Jungle*. Teacher Created Materials.

_____. 2013. *Survival! Ocean*. Teacher Created Materials.

Society of Primitive Technology. 2014. *Bulletin of Primitive Technology*. Rexburg, ID: Society of Primitive Technology.

Websites

Lundin, Cody. *Aboriginal Living Skills School*. http://www.codylundin.com/.

Noble, Christian. *Master Woodsman*. http://masterwoodsman.com/.

Nyerges, Christopher. *School of Self-Reliance*. http://www.christophernyerges.com/.

Magazines

Richie, Charlie (ed.). *The Backwoodsman*. Backwoodsman Magazine.